# AI REVOLUTION

## PROFITABLE OPPORTUNITIES IN THE AGE OF ARTIFICIAL INTELLIGENCE

AN E-BOOK BY

CARL P.

TABLE OF
# CONTENTS

# Introduction

Artificial intelligence, or AI, is changing the world. From self-driving cars to personalized recommendations on streaming services, AI is transforming the way we live, work, and play. And as the technology advances, the opportunities for making money with AI are only growing.

In this ebook, we will explore the AI revolution and the many ways to profit from it. We'll start by providing an overview of AI and its potential, before diving into the many opportunities available to investors, entrepreneurs, and job seekers. Along the way, we'll also examine the challenges and risks of working with AI, and consider the ethical considerations of this rapidly evolving technology.

# CHAPTER I

What is AI?

Before we can explore the opportunities of AI, it's important to understand what it is and how it works. AI refers to machines that can perform tasks that would normally require human intelligence, such as understanding natural language, recognizing patterns, and making decisions. These machines use algorithms and data to learn and improve over time, becoming more accurate and efficient as they go. There are many different forms of AI, each with its own strengths and limitations. Some AI systems are based on rules, in which a set of instructions are used to determine how the system should behave. Other systems are based on machine learning, in which the system is trained on data and learns to recognize patterns on its own. Deep learning is a subset of machine learning that uses neural networks, which are designed to mimic the structure of the human brain.

*The potential of AI is immense. It has already been used to improve healthcare, increase efficiency in manufacturing, and enhance security in airports. As the technology advances, it is likely to be used in even more ways, from predicting weather patterns to diagnosing diseases.*

*Opportunities for Investors*

*Investors have many opportunities to profit from the AI revolution. One of the most straightforward is to invest in AI-focused companies. These include companies that develop AI technology, as well as those that use AI to improve their products and services. Examples of AI-focused companies include Nvidia, Alphabet, and IBM.*

*Investors can also invest in venture capital funds that specialize in AI. These funds provide capital to startups that are developing AI technology or using it to disrupt existing industries. Examples of AI-focused venture capital firms include Sequoia Capital and Andreessen Horowitz.*

Investing in AI can be lucrative, but it is not without risks. One challenge is that the technology is still in its early stages, and it can be difficult to predict which companies and technologies will succeed. Investors also need to be aware of the potential for bias in AI algorithms, which can have negative consequences for both individuals and society as a whole.

Opportunities for Entrepreneurs

Entrepreneurs also have many opportunities to profit from the AI revolution. One option is to start an AI-focused company. This could involve developing new AI technology, or using existing technology to disrupt an industry. Examples of successful AI startups include UiPath, which provides robotic process automation, and Vicarious, which is working on developing AI that can think like a human.

IAnother option for entrepreneurs is to develop AI-powered products or services. For example, an entrepreneur could create an AI-powered chatbot that helps customers find products on an e-commerce website, or an AI-powered music recommendation service. These types of products and services can be highly lucrative if they provide value to customers and solve real problems.

Starting an AI business can be challenging, however. One issue is that the technology is complex and requires specialized expertise. Entrepreneurs also need to be aware of the ethical considerations of working with AI.

# Investing in AI: Risks and Rewards

Artificial intelligence (AI) is rapidly becoming one of the most important technologies of our time. From self-driving cars to virtual assistants, AI is changing the way we live and work. As the AI revolution continues, many investors are looking for ways to profit from this emerging market. In this chapter, we'll explore the various ways to invest in AI, including stocks, venture capital, and AI-focused funds. We'll also provide an analysis of the risks and rewards of each option, as well as tips for evaluating AI investment opportunities.

# CHAPTER II

*Investing in AI Stocks*

*One of the most direct ways to invest in AI is by buying shares of companies that are developing or using AI technology. Some of the largest tech companies in the world, such as Google, Microsoft, and Amazon, are heavily invested in AI research and development. Other companies, such as NVIDIA and AMD, specialize in designing AI-specific computer hardware. When investing in AI stocks, it's important to consider the company's financial health, its competitive position in the market, and its AI-related patents and research.*

*However, investing in individual stocks can be risky, particularly in emerging markets like AI. Some companies may fail to deliver on their promises, or may be overtaken by competitors with better technology. In addition, AI stocks may be subject to regulatory risks, particularly as concerns around data privacy and bias continue to grow.*

*Investing in AI Venture Capital*

*Another option for investing in AI is through venture capital (VC) funds. These funds invest in startups that are developing AI technology, often at an early stage. VC investors can provide not just funding, but also strategic guidance and industry connections to help these startups grow and succeed.*

*VC funds can provide a way to invest in a diverse portfolio of AI startups, spreading the risk of individual company failures. However, VC investing is also high-risk, as the majority of startups fail. Investors should carefully evaluate the experience and track record of the VC fund's management team, as well as the startups in which the fund is investing.*

Investing in AI-Focused Funds

For investors who want to invest in AI but don't want to pick individual stocks or startups, AI-focused funds may be a good option. These funds typically invest in a variety of AI-related companies, including software, hardware, and services providers. Some examples of AI-focused funds include the Global X Robotics & Artificial Intelligence ETF and the AI Powered Equity ETF.

AI-focused funds can provide diversification and professional management, but they may also be subject to higher fees and lower returns than investing in individual stocks or VC funds.

Evaluating AI Investment Opportunities

When evaluating AI investment opportunities, it's important to consider the potential risks and rewards of each option, as well as your own investment goals and risk tolerance.

It's also important to consider the ethical and social implications of investing in AI, particularly as concerns around bias, job displacement, and privacy continue to grow. Investors should look for companies or funds that have a strong track record of innovation and leadership in the AI space, as well as a solid financial foundation. They should also consider the competitive landscape and the potential for disruptive technologies or market shifts.

Conclusion

Investing in AI can be a high-risk, high-reward proposition. As the AI revolution continues, it's likely that we'll see new investment opportunities emerge, as well as new risks and challenges. By carefully evaluating potential investment opportunities and staying up-to-date on the latest AI trends and developments, investors can position themselves for success in this exciting and rapidly-evolving market.

# Starting an AI Business: Challenges and Opportunities

*Artificial intelligence (AI) is changing the world, and many entrepreneurs are eager to build successful businesses that leverage this powerful technology. In this chapter, we'll explore the process of starting an AI business, including developing an idea, building a team, and securing funding. We'll also provide case studies of successful AI startups, and offer advice on how to navigate the challenges of building an AI business.*

*Developing an AI Business Idea*

*The first step in starting an AI business is to develop a compelling idea that solves a real-world problem. This could involve using AI to automate repetitive tasks, improve decision-making, or enhance customer experiences. To develop an AI business idea, it's important to understand the latest trends and developments in the field, as well as the needs and pain points of your target market.*

*Building an AI Business Team*

*Once you have an idea for your AI business, the next step is to build a team of experts who can bring your vision to life. This may involve hiring data scientists, engineers, and software developers who have experience working with AI technologies. It's important to find team members who have both technical expertise and business acumen, as well as a passion for your mission.*

Securing Funding for Your AI Business

Building an AI business can be expensive, particularly in the early stages when you're developing your technology and building your team. To secure funding for your AI business, you'll need to develop a solid business plan that outlines your strategy, target market, and financial projections. You may also need to pitch your idea to investors or apply for grants or other funding opportunities.

Case Studies of Successful AI Startups

To help illustrate the process of building a successful AI business, let's look at some case studies of companies that have done it right.

One example is DeepMind, a London-based AI startup that was acquired by Google in 2015 for a reported $600 million. DeepMind's AI technology is focused on solving complex problems in areas like healthcare and climate change.

The company was founded by a group of experts in machine learning and neuroscience, and was able to secure early funding from investors like Elon Musk and Peter Thiel.

Another successful AI startup is Cognitivescale, a Texas-based company that provides AI-powered software solutions for businesses. Cognitivescale was founded by a team of experts in machine learning and enterprise software, and has raised over $70 million in funding to date. Navigating the Challenges of Building an AI Business Building an AI business can be challenging, particularly in a rapidly-evolving market where technology and competition are constantly changing. Some of the key challenges you may face when building an AI business include:

- Recruiting and retaining top talent in a competitive job market.

- Building scalable and reliable infrastructure for AI applications.

- Navigating complex regulatory environments and ethical considerations around data privacy and bias.

- Keeping up with rapidly-changing technologies and staying ahead of the competition.

To navigate these challenges, it's important to stay informed and connected with the broader AI community. Attend conferences, join industry groups, and connect with experts in your field to stay up-to-date on the latest trends and best practices.

Conclusion

Starting an AI business can be a challenging but rewarding endeavor. By developing a compelling idea, building a talented team, and securing funding, entrepreneurs can position themselves for success in this exciting and rapidly-evolving market. By learning from the successes and challenges of other AI startups, and staying up-to-date on the latest trends and developments, entrepreneurs can build innovative and impactful AI businesses that change the world.

# Careers in AI: Opportunities and Skills

Artificial intelligence (AI) is transforming industries and creating new opportunities for professionals with the right skills and expertise. In this chapter, we'll explore the various career opportunities in AI, including data science, machine learning engineering, and AI research. We'll provide an overview of the skills and qualifications needed for each role, as well as tips for finding and applying for AI jobs.

Data Science

Data science is a critical part of AI, as it involves using statistical methods and machine learning algorithms to analyze large amounts of data and extract insights. Data scientists work with structured and unstructured data from a variety of sources, including social media, sensors, and customer interactions. They use tools like Python, R, and SQL to analyze data, create models, and develop algorithms.

To become a data scientist, you typically need a degree in a field like computer science, statistics, or mathematics, as well as experience with machine learning and statistical modeling. Some of the key skills required for data science include:

- Proficiency in programming languages like Python and R.
- Knowledge of statistics and machine learning algorithms.
- Familiarity with data visualization tools like Tableau and Power BI.
- Strong communication skills to explain data findings to non-technical stakeholders.

Machine Learning Engineering

Machine learning engineering involves building and implementing algorithms that enable machines to learn and improve on their own. Machine learning engineers work with data scientists to build models that can be deployed in real-world applications. They use tools like TensorFlow, PyTorch, and Apache Spark to develop and train machine learning models.

- To become a machine learning engineer, you typically need a degree in computer science, software engineering, or a related field. Some of the key skills required for machine learning engineering include:

- Proficiency in programming languages like Python, Java, and C++.

- Familiarity with machine learning frameworks like TensorFlow and PyTorch.

- Understanding of algorithms and data structures.

- Strong problem-solving and critical thinking skills.

AI Research

AI research involves developing new techniques and algorithms for machine learning and other AI applications. AI researchers work in academia, industry, and government, and may focus on areas like natural language processing, computer vision, or robotics. They use tools like Python, TensorFlow, and MATLAB to develop and test new algorithms.

To become an AI researcher, you typically need a PhD in a field like computer science, mathematics, or electrical engineering. Some of the key skills required for AI research include:

- In-depth knowledge of machine learning algorithms and techniques.
- Strong programming skills in languages like Python, Java, and C++.
- Familiarity with tools like TensorFlow, PyTorch, and MATLAB.
- Experience conducting research and publishing papers in academic journals.

Finding and Applying for AI Jobs

If you're interested in pursuing a career in AI, there are several ways to find and apply for jobs in the field. Some strategies include:

- Networking with professionals in the AI community, both online and in-person.
- Attending AI-focused job fairs and career events.
- Using job search websites like Indeed, Glassdoor, and LinkedIn to search for AI jobs.
- Applying directly to companies that are known for their AI expertise, like Google, Amazon, and Microsoft.

When applying for AI jobs, it's important to tailor your resume and cover letter to highlight your relevant skills and experience. Be sure to mention any specific AI projects you've worked on, as well as any relevant coursework or certifications. During interviews, be prepared to discuss your technical skills and problem-solving abilities, as well as your passion for AI and its potential impact.

Conclusion

AI is a rapidly-growing field with a wide range of career opportunities for professionals with the right skills and expertise.

# Chapter 5: Ethics and the Future of AI

As AI becomes more advanced and integrated into our daily lives, it raises important ethical considerations that must be addressed. In this chapter, we will explore some of the most pressing ethical issues surrounding AI, as well as the potential future of this technology.

The Ethics of AI One of the primary concerns about AI is the potential for bias. AI systems are only as unbiased as the data they are trained on, and if that data is biased in any way, the system will be biased as well. This can have serious consequences, particularly in areas such as hiring, lending, and criminal justice. It is important that AI developers and users take steps to identify and mitigate bias in these systems.

# CHAPTER V

Privacy is another major ethical concern when it comes to AI. As AI systems become more advanced, they may be able to collect and analyze vast amounts of data about individuals without their knowledge or consent. This raises questions about how that data will be used, and whether individuals have the right to control their own data.

Finally, the increasing automation of jobs through AI has the potential to leave many people without work, leading to economic and social upheaval. It is important that we consider the potential impacts of AI on employment and take steps to mitigate any negative effects.

The Future of AI Despite these ethical concerns, the potential of AI to transform our lives is immense. In the healthcare industry, AI has the potential to improve diagnostics and treatment, leading to better outcomes for patients. In the transportation industry, self-driving cars could reduce accidents and improve traffic flow.

And in the business world, AI has the potential to automate mundane tasks, freeing up employees to focus on more creative and strategic work.

However, there are also concerns about the future of AI. Some experts have warned about the potential for AI to become a threat to humanity, either through deliberate misuse or unintended consequences. It is important that we continue to develop and improve AI in a responsible way, taking into account the potential risks and benefits of this technology.

Preparing for the Future As AI continues to evolve, it is important that individuals and businesses alike prepare for the changes that lie ahead. This means developing the skills and knowledge needed to work with AI systems, as well as considering the ethical implications of these technologies. Businesses that are able to successfully integrate AI into their operations will have a competitive advantage in the years to come.

# CHAPTER V

Conclusion AI is a transformative technology with the potential to revolutionize our world in countless ways. However, it also raises important ethical concerns that must be addressed. By understanding these issues and preparing for the future, we can ensure that we make the most of the opportunities presented by the AI revolution while minimizing any potential risks.

www.ingramcontent.com/pod-product-compliance
Lightning Source LLC
Chambersburg PA
CBHW071147220526
45467CB00015B/2092